Emile-Michel O. L. NDJADIANDJA

OVERVIEW AND OBJECTIVES FOR PHYSICAL SCIENCES

Emile-Michel O. L. NDJADIANDJA

OVERVIEW AND OBJECTIVES FOR PHYSICAL SCIENCES

In the First Year of Scientific Humanities Classes

ScienciaScripts

Imprint
Any brand names and product names mentioned in this book are subject to trademark, brand or patent protection and are trademarks or registered trademarks of their respective holders. The use of brand names, product names, common names, trade names, product descriptions etc. even without a particular marking in this work is in no way to be construed to mean that such names may be regarded as unrestricted in respect of trademark and brand protection legislation and could thus be used by anyone.

Cover image: www.ingimage.com

This book is a translation from the original published under ISBN 978-620-3-41430-1.

Publisher:
Sciencia Scripts
is a trademark of
Dodo Books Indian Ocean Ltd., member of the OmniScriptum S.R.L Publishing group
str. A.Russo 15, of. 61, Chisinau-2068, Republic of Moldova Europe
Printed at: see last page
ISBN: 978-620-3-66377-8

PREFACE

In order to understand the principles of physics, it is not enough to simply give lectures and solve numerical exercises, no matter how good they may be; it is essential to progress in the theoretical and epistemological study of the subject.

Since students have a lot of difficulties with numerical applications and solving physics problems, the author proposes a book entitled "***Overview and objectives for physical sciences***" in general secondary education, combing through the exercise of solving these problems.

Isn't it contradictory to put the unit of measurement at the end of a long operation, in relation to the resolution of a physics exercise?

The essential principle to which it first strives is to invite the learner to use the essential tool to do so: the "stubborn" use of units of measurement, at all times, from the announcement of the problem to the result, including any necessary transformations. What we summarize by the term " DIFORAN "[= Data, Unknown(s), FORmule and Digital Application, of the remainder sufficiently exploited] step.

For this, Professor ONEMA, didactician of physics has taken care to exploit the works of these predecessors as competent and critical: Basquin, A; Bribosia, A; Faucher, R; Fontaine, G; Jodogne, J; PIRA, E and Mpanda, M and Luyindula, M.

What is this relevant average author that would allow the learner to excel in solving physics number problems?

Indeed, the norms of writing and typography of texts are required at all times; among other things, avoid fractional format for units of measurement (kg/m3; m/s2; ...), correctly separate the compound units of measurement (m. kgf or m kgf), as stipulated in the current classics.

Our most ardent wish: That Mr. ONEMA continues on the same path for the other classes of general education. Our encouragement and support for this oil task that is inexorably expanding.

Kinshasa, March 14, 2021 José-Em- MATA TOMBO

Discipline Didactician and Trainer

FOREWORD

In order to improve the quality of physics teaching in the Democratic Republic of Congo (D.R.C.), we are making available to teachers and students of the first year of Scientific Humanities the first volume of the manual "Overview and objectives for physical sciences".

This textbook contains subjects that meet the demands of students and teachers, in accordance with the National Physics Program in force for the first year of the Scientific Humanities. To this end, in addition to some new notions added for a wider knowledge of the course, you will find application exercises at the end of each subject.

We are convinced that this manual is a support for our teaching in difficult conditions. It will help our students to acquire structured knowledge about physics and will reduce the difficulties of the teachers, who are sometimes struggling with the lack of appropriate documentation.

We thank Mr. Robert Nyami Bwapangu, a physics teacher at the Masina Pedagogical and Scientific Institute, for reading the manuscript of this book.

Our early thanks go especially to all those who will help us to improve this book. Criticism, remarks, advice, suggestions from all those competent in the field who want to help and encourage us will be received with great respect and are of the highest quality.

The author

HISTORICAL OVERVIEW OF PHYSICS

Physics, as an experimental science and objective to scientific progress, is the relationship between man and the phenomena of nature. It draws its origins from the immediate needs of the society in particular: to cultivate, to raise, to write, to communicate, to locate, to measure, to control, ... in short to work.

A knowledgeable person in the field of physics has access to the scientific evolution of Copernicus, who avoided the religious conception in order not to be blocked to make researches to the point of discovering the phenomena by observation and experimentation in order to govern the laws.

The evolution of physics is closely related to the foreseeable technological uses that are increasingly dependent on the economic and political factors of the country. Scientific evolution is related to man exploiting physics to find technology.

Therefore, physics is a science of interest linked to technology that is at the service of man, as summarized below:

Physics ↘

Men ←Technology

ABBREVIATIONS AND ACRONYMS

A: ampere

Ah: ampere hour

C : coulomb

d.d.p : difference in potentilla

OCC: Congolese Control Office

D.R.C.: Democratic Republic of Congo

SI: International System of Units

t : time

U.P.N : National Pedagogical University

µ : moment of a force

TABLE OF CONTENTS

INTRODUCTION

Physics is an objective and experimental science that studies reversible phenomena. It has its origins in the immediate needs of society.

Physics uses observational and experimental methods to determine the laws that govern reversible phenomena.

Physics is related to technology based on the invention of tools used in industries by man and it studies the phenomena that do not deeply alter the intimate nature of bodies, including the movements of bodies, the change of physical states of matter, the reflection of light, the refraction of light, the free fall of bodies, the radioactive decay, the expansion of bodies.

Physics differs from chemistry in phenomena and teaching methodology. It uses technology to govern the laws (principles).

Physics is useful for mankind that obeys the recommendations of mechanics, electricity and optics.

N.B. Reversible phenomena are events that occur over a period of time and require a measure of magnitude to govern the laws.

CHAPTER I: CONCEPTS OF METROLOGY

Objectives

Knowledge : to note the importance of metrology

Know-how: to restore the definition of metrology from the information received.

I.1 DEFINITIONS

- **Metrology** is the science of measurement. It studies all the variables involved: the characteristics of the object to be measured, the measuring instrument, the method of measurement, the influence of the environment, the determination of the unit of measurement and of a measurement system that is consistent, universal and permanent, and the constitution of the calibration chain allowing maximum accuracy. Metrologists are particularly interested in determining the causes of error in measurements, in order to discover ways to avoid them, or at least to reduce them.
- **Measurement** is the action of evaluating a quantity according to its relationship with a quantity of the same kind, taken as unit and as reference, quantity, dimension thus evaluated.
- **To measure a quantity is** to compare this quantity to another quantity taken as unit.
- **Measuring is** the action of measuring by a direct and concrete process.
- **Measurand** is any quantity to be measured such as length, mass, time, etc.

I.2 TYPES

We distinguish:

- Fundamental or scientific metrology aims at creating, developing and maintaining reference standards. This metrology is used in the educational systems of the world,
- The technical or industrial metrology: it allows to guarantee often the framework of quality control linked to a system, the quality management. Economic and industrial operators use this metrology,
- Legal metrology: this is related to measurement activities to which regulatory requirements apply. It is used by control and conformity to standards organizations, such as the Office Congolais de Contrôle.

CHAPTER II: SYSTEM OF UNITS OF MEASUREMENT

Objectives

Knowledge: name the fundamental quantities

Know-how: indicate the instrument used to measure length, mass, time, ... calculate the dimensions of a square, a rectangle.

II.1 NOTION ON QUANTITIES

1. Size

A quantity is anything that is measured by a measuring instrument. Magnitudes are variable from one way to another and they are comparable to the unit considered.

Quantities require measurement to determine the laws that govern reversible phenomena. The quantities correspond to length, mass, time, temperature, force, weight, pressure, surface, volume, acceleration, speed, capacity, work, energy, etc.

The fundamental quantities are: length, mass and time and they require a system of units MKS, CGS and MKfS.

The derived quantities correspond to velocity, acceleration, force, weight, volume, density, area, capacity, work, energy, temperature, heat quantity, electric charge, etc. They require a system of units MKSA which constitutes an international system (SI).

The vector quantities correspond to force, acceleration, velocity and weight.

N.B.: To measure a quantity is to compare this quantity to another quantity of the same kind taken as a unit.

MKS stands for meters, kilograms and seconds,

MKSA stands for meter, kilogram, second and ampere,

CGS stands for centimeter, gram and second.

2. Length

Length is a distance between the extremes of a body and corresponds to the distance (space) traveled by a body, the height (altitude) reached by a body, the elongation of a body's movement and the displacement of a body.

Length is measured in meters as the primary unit and is also measured in multiples and sub-multiples of the meter, including :

1AL (light-year) = 9.5×10^{12} km

1 AU (Astronomical Unit) = 1.5×10^{8} km

1 km = 1000 m = 10^{3} m

1 hm = 100 m = 10^{2} m

1 dam = 10 m

1 dm = 0.1 m = 10^{-1} m

1 cm = 0.01 m = 10^{-2} m

1 mm = 0.001 m = 10^{-3} m

1 μ (micron) = 0.000001 m = 10^{-6} m

1 nm (nanometer) = 10^{-9} m

1 Å (Angstrom) = 10^{-10} m

1 pm (pico meter) or 1micron micron (μμ) = 10^{-12} m

1 fermi = 10^{-15} m

N.B.: The Light Year is a distance traveled by light in 1 year in the inter-satellite space. The Astronomical Unit is an average distance that separates the cold stars from the sun. The synoptic table of the length measurement is indispensable.

Application exercises

Convert the following lengths to millimeters (mm):

1) 3,5 Å
2) 0,52 μ
3) 3,5 km
4) 0.01 μμ
5) 2,5 fermis
6) 0,02 Å
7) 0.002 AL

Answers:

1) 3.5 Å = 3.5 x 10^{-10} m = 3.5 x 10^{-7} mm = 35 x 10^{-8} mm
2) 0.52 μ = 52 x 10^{-5} mm
3) 3.5 km = 35 x.10^{5} mm
4) 0, 01 μμ = 10^{-11} mm
5) 2,5 fermis = 25 x10^{-12} mm
6) 0.02 Å = 2 x10^{-9} mm
7) 0.002 AL = 19 x10^{16} mm

3. Mass

The mass is a quantity of the concentrated matter that forms a body. It depends on the number of molecules and the position of the body.

Mass is measured in kilograms as the fundamental unit and is also measured in multiples and sub-multiples of kilograms, including :

1 ton = 1000 kg = 10^{3} kg

1 quintarre = 100 kg = 10^{2} kg

1 hg = 0.1 kg = 10^{-1} kg

1 dag = 0,1 kg = 10^{-2} kg

1 g = 0.001 kg = 10^{-3} kg

1 dg= 0, 0001 kg =10^{-4} kg

1 cg= 0, 00001 kg = 10^{-5} kg

1 mg = 0.000001 kg = 10^{-6} kg

N.B.: The synoptic table of the mass measurement is necessary.

The scale is a device used to measure mass.

Application exercises

Convert the following masses to kilograms (kg):

1) 252 x 10^{-2} t
2) 25 g
3) 0.152 mg
4) 16 cg

Answers

1) 252 x 10^{-2} t = 252 x 10^{-2} x 10^{3} kg = 2520 kg
2) 25 g = 25 x 10^{-3} kg
3) 0.152 mg = 152 x 10^{-9} kg
4) 16 cg = 16 x 10^{-5} kg

Ref : J. Jodogne ; Physique 3 ième : A De Boeck, 1970, p 25.

4. Time

Time is a period of time that flows uniformly without stopping, without accelerating, and without slowing down during the day or night.

Time is measured in seconds as the primary unit and is also measured in multiples and sub-multiples of the second, including:

1 day = 86400 s

1 hour = 3600 s

1 minute = 60 s

1 ns = 10^{-9} s

1 ps = 10^{-12} s

1 third = 17 x 10^{-3} s

N.B.: 1 year = 365 days and 6 hours

1 century = 100 years.

Application exercises

1). Convert 40,000 seconds and 2,000 minutes to hours.

2). Convert 2h25min 32 sec and 3h20min into seconds.

Answers

1).

a) 40 000 s = 11 h 06 min 40 s.
b) 2 000 min = 33 h 20 min.

2).

a) 8,732 seconds.
b) 12,000 seconds.

Ref : **J. Jodogne ; Physique 3 ième ; A De Bock, 1970, p.25.**

5. Temperature

Temperature is a characteristic of the heat and weight given off by a body. It corresponds to the tactile sensation of a hot or cold body.

Temperature is measured in Celsius (°C) in the Celsius scale, in Kelvin (°K) in the Kelvin scale, in Fahrenheit (°F) in the Fahrenheit scale or in Réaumur in the Réaumur scale.

The temperature indicated in the boiling steam is 100°C and that indicated in the melting ice is 0°C. The temperature limits for water heated by sterilization and for cooking in the autoclave vary from 120 to 125°C.

The boiling temperature rises when there is an increase in pressure caused by the boiling water. The melting temperature reaches 2000 °C for the transformation of the iron ore into cast iron and the gangue into slag in the blast furnace. The temperature is measured in the hot body and in the cold body. The boiling temperature is lowered when there is a decrease in pressure caused by solidification. The melting temperature remains constant during the whole melting period. The solidification temperature remains constant during the whole solidification period.

The melting temperature corresponds to the solidification temperature. The temperature limits of a sick person vary from 37 to 42°C in the pre-noon and 37.5 to 42°C in the afternoon. The limiting temperatures of a normal person vary from 36 to 36.9 ° C in the pre-noon and from 36 to 37 ° C in the afternoon.

The temperature is accurately indicated on the expansion of mercury in the glass envelope of a thermometer.

N.B.: The thermometer is based on the expansion of mercury in the glass envelope.

100 °C = 80 °R = 180 °F

0 °C= 0 °R = 32 °F

0 °C = 273 °K.

II.2 NOTIONS ON ERRORS

Objectives

Knowledge: name the different types of measurement errors

Know-how: solve the exercises on measurement errors

1. Measurement of quantities

The measure of a magnitude is a concrete number consisting of the digits followed by the chosen unit and it gives the idea to look for how many times the magnitude contains the considered unit.

The measure of a quantity is composed of the significant digits except zero preceding the first digit different from zero. The measure of 524 cm is composed of significant figures 5, 2 and 4.

The measurement of 0.0524 has significant figures 5, 2, 4 and the two zeros are not significant figures. The measurement of 52.4 cm is extended to the nearest 0.1 cm. The measure of 52.40 cm is pushed back to the nearest 0.01cm.

Measurements of magnitude are accurately determined by an accurate instrument.

Measurements of magnitude are determined with systematic error by a defective instrument.

Accurately determined measurements of magnitude are sought with care, with all seriousness, with all attention and not in a simple way.

Measurements of magnitudes determined with systematic error (which occurs with a certain order and method) are serious, but in order to eliminate these errors, they must be carefully sought.

Measurements of magnitude are determined with accidental error by the observer and by the instrument.

Measurements of magnitude determined with accidental error are due to the imperfection of the observer, notably bad hearing, bad sight, slow reflex, fatigue, etc., to the imperfection of the instrument, notably the defect of sensitivity, the defect of fidelity, etc.

Measurements of magnitude shown with accidental error depend on the position of the observer.

N.B.: the accidental error is random and therefore unavoidable, the accidental error depends on the position of the observer corresponds to the parallax error.

The systematic error always comes from the same cause which is the inherent defect of the instrument. 2 km =2000 m, the zeros are not significant figures except 2. 98 x 87 = 8526 km ^{2}, the significant figures to keep are 2 and 6 to obtain 8500 m ^{2}.

A meter that is too short always gives too high numbers, a watch that is moving forward always gives too high numbers and a scale indicates a very low mass when the mass of the object is greater than the mass marked on it.

2. Absolute error

The absolute error is a difference between the exact measurement and the estimated measurement of the quantity.

The absolute error committed on the measurement of magnitude is defined by :

ΔE = E2 - E1

where ΔE : absolute error in SI

E $_2$: measure assessed in SI

E $_1$: exact measurement in SI.

N.B.: The absolute uncertainty theorem requires:

X= a+b+c; becomes Δx = Δa+Δb+ Δc

The result of the limit measurements is between lower and upper value ($v_i<v<v_s$). The exact measurement is still not known.

Application exercises

1. A metal bar is between the 80cm and 81cm lines of the metre divided into centimetres. Calculate the absolute error made on the measurement of the length of a bar.

Answer: $\Delta l \, l = \frac{80+81}{2} = 80{,}5cm$ **= 80.5 - 80 = 0.5 cm**

So Δl = 80 - 80.5 = 0.5 cm

2. A book cover measures 18.89 cm long and 18.85 cm wide. Calculate the absolute error in the length of a book cover. **Answer: 0.02 cm.**

3. A rectangle whose sides measure 4.8 cm and 7.5 cm with an absolute error of 0.005 cm for each measurement ^{2}. Express the result of the boundary measurements of the area of a rectangle.

Answer: 35.99 < 36 < 36.01 cm 2

4. A length of a bench measures (1273 ± 1) cm. Find the resultant of the limiting measurements of the length. **Answer: 1272 < 1273 < 1274 cm.**

5. The sides of a triangle measure 3 cm, 4 cm and 5 cm with each measurement having an absolute error of 0.1 cm. Calculate the perimeter of a triangle. **Answer: P = (12± 0.3) cm.**

6. A rectangle measures a length of (15±0.1) m and a width of (10.0±0.1) m. Calculate the area of the triangle. **Answer: (150.0 ±0.2) m ^{2}.**

Ref : E. PIRA; Energy; Kinshasa 1973, p.8.

3. Relative error

The relative error is a ratio that exists between the absolute error committed on the quantity measurement and the experimental measurement of the quantity is defined by :

$$E_r = \frac{\Delta E}{E}$$

where E_r : relative error in %.

ΔE : absolute error in SI

E : experimental measurement in SI.

N.B.: The accuracy of a measurement of quantity is greater the smaller the relative error.

On a measurement of E^n, the relative error $E_r = Ne_r$ on E.

Application exercises

1. A length of a pencil measures (15.6±0.1) cm. Calculate the relative error made on the length of a pencil.

Response: $E_r = \frac{0,1}{15,6}$

E_r **= 0.0064**

E_r **= 0.64%.**

2. A length measurement is close to (243.5±0.02) cm. Calculate the relative error on the length. **Answer: 0.0082 %.**

3. A length is indicated by measurements of 0.86±0.01) m; (9.3±0.1) m; (2.14± 0.01) cm and (80± 0.2) m. Which of these measurements is the best?

Answer: This is the measurement of 2.14 ± 0.01) cm.

4. The sides of a rectangle are measured at 4.8 and 7.5 cm at $\frac{5}{1000}$. Calculate the relative error made on the area of the rectangle. **Answer: 0.028%.**

5. A side of a perfect cube measures 2.4 cm at $\frac{1}{100}$. Calculate the relative error on the volume of a perfect cube. **Answer: 0.072%.**

6. A side of a perfect cube measures 2.4 cm at $\frac{1}{100}$. Calculate the relative error in the volume of a perfect cube. **Answer: (14±0.03) cm3.**

7. A volume of a perfect cube measures 27 cm ^{3} at $\frac{5}{100}$. Calculate the relative error made on the volume of a perfect cube. **Answer: 0.19%.**

8. A volume of a perfect cube measures 27 cm ^{3} at $\frac{5}{100}$. Calculate the relative error made on the side of a perfect cube. **Answer: 1.7%.**

9. A height of a cylinder of 10 cm ^{2} measures 15.6 cm. Calculate the relative error in the volume of a cylinder measured at $\frac{5}{100}$. **Answer: 0.032%.**

E. PIRA; Energy; Kinshasa, 1973, p 9

CHAPTER III : KINEMATICS

Objectives

Knowledge: Name the effects of force

Know-how: solve the kinematics exercises

III.1 ACTION OF A FORCE

1. Driving force

The driving force is a capable external cause that moves its point of application. The driving force is assumed to be constant in direction, sense and intensity defined by the fundamental law of dynamics:

$$\mathbf{F = m.a = m.v/t}$$

where F : driving force in N or kgf

m : mass of the body in kg

a: acceleration of motion in $m.s^{-2}$

v : speed in meters per second ($m.s^{-1}$)

t : time in seconds (s).

N.B.: Any body that is not subjected to the action of a force retains its state of rest or motion. An inert body cannot set itself in motion. A body in a state of inertia cannot of itself change its state of rest or motion.

At zero force, at zero acceleration. At constant force, at constant acceleration.

1 Kgf = 9.81 N.

Application exercises

1. A body of two kilograms is subjected to a driving force of five newtons in 20 seconds. Calculate the speed of the body.

Answer

Data	Request	Formula
m= 2 kg F = 5 N t= 20 s	v = ? in m.s $^{-1}$	$F = m\,a = m\frac{v}{t}$
Numerical application		
$F = m\frac{v}{t}$; $v = \frac{F\,x\,t}{m}$ $v = \frac{5\,x\,20}{2} = 50m.s^{-1}$		

2. A 150 ton locomotive develops a force of 6.10 ^{4} Newtons in 50 seconds. Calculate the speed of the locomotive. **Answer: 20 m.s $^{-1}$**.

3. An automobile of 1200 kilograms has a speed of 90 km/h after 20 seconds. Calculate the force exerted by the engine. **Answer: 1800N**.

3. A 6 ton truck is driven at a speed of 36km per hour after 20 seconds. Calculate the force exerted by the engine. **Answer: 3.10 ^{3} N**

4. A lead ball of 5 grams has a speed of 3 x 10 ^{8} m.s $^{-1}$. Calculate the force exerted on the ball after one second. **Answer: 15.10 ^{5} N.**

5. A 2 kg mobile is subjected to a constant force of 5 N. Calculate the acceleration of the mobile. **Answer: 2.5 m.s $^{-2}$.**

Ref : J. Jodogne ; physique 3 ^{ième} ; A De Boeck 1970, p 79.

2. Gravity

Gravity is a force of attraction exerted on a celestial body moving vertically from top to bottom.

Gravity is the weight of the body:

P= m.g

where P: body weight in N or in Kilograms-weight (kgp),

m : mass of the body in kg,

g : acceleration of gravity in m.s $^{-2}$.

N.B.: In Paris, the acceleration of gravity g = 9,81m.s $^{-2}$ $\approx$ 10 m.s $^{-2}$.

The weight of the body increases with latitude and decreases with altitude.

The weight of a body is zero when there is no gravity. The acceleration of gravity decreases when the body rises in altitude. The acceleration of gravity pushed to 0,5m.s $^{-2}$ is rounded by excess and that pushed to 0,4 m.s $^{-2}$ is rounded by default.

Application exercises

1. An object of 50 kg is located at an altitude of o m (g = 9,81 m.s $^{-2}$). Calculate the weight of the body.

Answer

Data	Request	Formula
m = 50 kg g = 9.81 m.s $^{-2}$	P = ? in N	P = m g
Numerical application		
P = 50 x 9.81 P = 490.5 N		

1. An object located at the altitude of 1 km (g = 9.807 m.s $^{-2}$) weighs 19.62 kgp. Calculate the mass of the object. **Answer: 2 kg.**
2. A body on the moon in the environment where g = 1.67 m.s $^{-2}$ weighs 17.7N. Calculate the mass of the body. **Answer: 10 kg**.
3. A body of 20 kg rises in altitude of 10 km at the level of the equator (g = 9,779 m.s $^{-2}$). Calculate the weight of this body. **Answer : 195,58 N.**
4. A body of 20 kilograms is in the vicinity of the earth at the altitude of 80° $^{-2}$). Calculate the weight of the body.

   Answer. 196,6 N.
5. An object of 1 kg is mentioned in Kinshasa where g = 9.7806 m.s $^{-2}$ and in Paris where g = 9.80943 m.s $^{-2}$.

a) Calculate the weight of the above body in Kinshasa. **Answer: 9.8 N.**

b) Calculate the weight of the above mentioned body in Paris. **Answer: 9.81 N**.

Ref: E. Pira; Energie; Kinshasa, 1970, p 53.

3. Work

Work is the capacity of the force that moves its point of application. The work done by a force exerted on a body is defined by :

$$T = F..d$$

where T : work in joules (J) or kgfm

F : force in N or in kgf (kg')

d: displacement (distance) in m.

N.B.: If the directions of the force and the displacement are the same, the work is driving; if the directions of the force and the displacement are opposite, the work is resisting and, if the directions of the force and the displacement are perpendicular, the work is null.

1 kgfm = 9.81 J

1 N.m = 1 J

Application exercises

1. A tractor pulling a cart exerts a force of 400 Newtons over a distance of 3 kilometers. Calculate the work of a force performed by a tractor.

Answer

Data	Request	Formula
F = 400 N d = 3 Km	T = ? in J	T = F.d
Resolution		
T = 400 x 3000 T = 12.10^{5} J		

2. A horse pulls a dump truck with a force of 40 newtons over a distance of 20 km. Calculate the work done by the horse. **Answer: 8.10^{7} J.**
3. A man moving a load exerts a force of 52.3 N over a distance of 23 m. Calculate the work done by the man. **Answer: 1202.9 J.**
4. A person lifts a bag weighing 50 kgp over a distance of 2 km. Calculate the work produced by this person. **Answer: 975.10^{3} J.**
5. A crane lifts a load of 40 kgp to a height of 20 m. Calculate the work done by the crane. **Answer: 7848 J.**

6. A coal merchant carrying 50 kg of coal climbs 6 floors of an urban building. The average height of a floor is 4 m.

a. Calculate the work done by the carrier transporting the coal.

   Answer: 11,760 J

b. What does the work done by the coalman depend on? **Answer: It depends on the average height of each floor and not on the shape of the stairs.**

Ref : J. Jodogne ; Physique 3 ième ; A de Boeck 1970, p 115.

4. Uniform Rectilinear Motion (URM)

Objectives

Knowledge: differentiate between the different movements

Know-how: calculate the space, speed, time and acceleration of a moving object

The uniform rectilinear motion is a translation of a body that travels through equal spaces in equal times. The uniform rectilinear motion of a body moving along a straight line at constant speed is defined by :

e = v.t

v = cste.

where v : speed of motion in $m.s^{-1}$

t : time in seconds (s)

e: distance covered in meters.

N.B.: The speed is a vector quantity which is related to the progression of the body.

Application exercises

1. A pedestrian walks a distance of 10.8 km at a speed of 5.4 km/h. Calculate the duration of the pedestrian's walk.

Answer

Data	Request	Formula
e = 10.8 km v = 5,4 km/h	t = ? in s	e = v t
Reasoning $t = \frac{e}{v}$ $t = \frac{10\,800}{1,5}$ t = 7 200 s		

1. A car drives at 30 $m.s^{-1}$ on a straight track in 10 minutes. Calculate the distance traveled by the car. Answer: 18.10^{3} m.

2. An automobile moves uniformly at a speed of 90 km per hour. Calculate the distance traveled by the automobile after 2 hours of motion. **Answer: 18.10^{4} m**

3. A car travels at a speed of 90 km per hour in 3 hours. Calculate the distance traveled by the car. **Answer: 270 kilometers**

4. A body travels a distance of 1500 m in 15 seconds. Calculate the speed of the body. **Answer: 100 $m.s^{-1}$.**

Ref: E. Pira; Energie, Kinshasa, 1973, p. 41.

5. Accelerated Uniform Rectilinear Motion (AURM)

Uniformly accelerated rectilinear motion is a translation of a body that moves along a straight line at constant acceleration. The uniformly accelerated rectilinear motion animated by a body that is related to the variation of velocity is defined by the system of time equations :

$$\mathbf{e = \frac{1}{2}a.t^{2}}$$

$$\mathbf{v = a.t}$$

where a : acceleration of motion in $m.s^{-2}$,

t : time of the movement in seconds,

e : space covered in meters.

N.B.: The acceleration of motion is a vector quantity which is related to the variation of speed of the motion of a body which progresses on the rectilinear trajectory.

Application exercises

1. A cyclist has a speed of 10 $m.s^{-1}$ in 4 min 10 s. Calculate the space covered by the cyclist.

Answer

Data	Request	Formula
$v = 10$ m.s $^{-1}$ t = 4 min 10 sec	e = ? in m	$e = \frac{1}{2}at^2$
Reasoning		
$V = a.t$; $a = \frac{v}{t}$ $e = \frac{1}{2}\frac{v}{t}t^2 = \frac{1}{2}vt$ $e = \frac{1}{2} \times 10 \times 250$ $e = 1250\,m$		

2. A body has a constant acceleration of 5 m/s ^{2} in 20 seconds. Calculate the space covered by the body.

Answer: 1000 meters.

3. A mobile is animated by an acceleration of 5 m.s $^{-2}$ at the end of the 6 th second of the movement. Calculate the speed of the mobile. **Answer : 30 m.s $^{-1}$.**

4. A moving object with an acceleration of 2 m.s $^{-2}$ travels a distance of 0.1 km. Calculate the speed reached by the mobile**. Answer : 20m. s $^{-1}$.**

5. A mobile launched at an acceleration of 2.5 m.s $^{-2}$ travels a distance of 2 km. Calculate the time after which the mobile is animated.

Answer: 40 s.

Ref: E. Pira; Energie; Kinshasa, 1973, p 42.

6. Free fall

Free fall is a vertical motion that corresponds to the translation of a body subjected to the force of gravity. The free fall animated by a body falling from top to bottom is defined by a system of equations :

$$\mathbf{h = \frac{1}{2} g.t^2}$$

$$\mathbf{v = g.t}$$

where h : height of fall in meters ;

g: acceleration of gravity in m.s $^{-2}$,

t : time of the movement in seconds,

v : speed of movement in m.s $^{-1}$.

N.B. : The acceleration of gravity is g = 9,81 m.s $^{-2}$. The direction of free fall is vertical. The direction of the free fall is up and down (direction).

Application exercises

1. A stone dropped from the top of a building takes 4 seconds to reach the ground. Calculate the height of the building.

Answer

Data	Request	Formula
t = 4 s	h = ? in m	$h = \frac{1}{2} g\ t^2$
Reasoning		
$h = \frac{1}{2} x\ 9{,}8\ m.s^{-2} x\ 16\ s^2$ h = 78.4 m		

2. A stone dropped from the top of a building takes 4 seconds to reach the ground. Calculate the velocity of the stone as it reaches the ground. **Answer: 39.2 m.s $^{-1}$.**

3. A stone falls in 5 seconds from the ground. Calculate the height of fall of the stone. **Answer: 122.5 meters.**

4. A stone is dropped at the height of 83 m to reach the ground.

   a) Calculate the time after which the stone falls. **Answer: 4.1** s.

   b) Calculate the velocity of the stone as it hits the ground. **Answer: 40.16 m.s $^{-1}$.**

5. A lead ball directed upward reaches a height of 122.5 m. Calculate the falling speed of the ball as it reaches the ground. **Answer: 49m.s $^{-1}$.**

6. In Kinshasa, the acceleration of gravity varies from 9.7806 to 9.8 m.s $^{-2}$. Calculate the relative error on the acceleration of gravity. **Answer: 0.2 %.**

III.2 STATIC EFFECTS OF A FORCE

Objectives

Knowledge: Name the effects of force

Know-how: calculate the moment of the force

1. Axis

The axle is a body that passes through the coupling. Axles hold bodies subject to a force that constitutes a moment:

μ : F.d

where μ : moment of a force in kgfm ;

F : force in newtons ;

d: distance in meters.

N.B.: The moment of a force corresponds to the work of a force.

Application exercises

1. James exerts a force of 5 kgf on each leg of a left-hand turn at a distance of 0.3 m. Calculate the moment of the applied force with respect to the axis.

Answer

<table>
<tr><td>Data</td><td>Request</td><td>Formula</td></tr>
<tr><td>F = 5 kgf
d = 0.3 m</td><td>μ = ? in kgfm</td><td>μ = F. d</td></tr>
<tr><td colspan="3">Reasoning</td></tr>
<tr><td colspan="3">μ = 0.3 x 5
μ = 1.5 Kgf. m</td></tr>
</table>

2. A force of 4 Kgf is applied at a point A of a body having an axis in 0 at the distance of 3 m. Calculate the moment of the applied force with respect to point 0. **Answer: 12 kgf. m.**
3. A force of 5 Kgf acts on a body rotating around an axis 0 at a distance of 6 m. Calculate the moment of the force exerted with respect to the axis. **Answer: 30 kgf. m.**

Ref : R. Basquin ; Mécanique expérimentale ; Paris, 1960, p 104.

2. Lever

The lever is a rigid bar that rotates around a support axis under the action of a driving force and a resisting force. Advantageous levers promote a driving force to overcome a resisting force in conditions such as:

$\mathbf{F_M B_M = F_R B_R}$

where:

F_M : driving force in Newtons,

F_R : resisting force in Newtons,

B_M : Lever arm of the FM in meters,

B_R : lever arm of the RF in meters.

N.B.: All interlocking levers are advantageous. No inter-motor levers are advantageous, notably the fire tongs and the pedal of a sewing machine. Inter-lever levers may or may not be advantageous.

Ref : J. Jodogne ; Physique 1, mécanique ; A De Boeck 1970, p 102.

Application exercises.

1. A stone of 2×10^3 N opposes 20 cm from the support axis which supports a bar. Calculate the driving force that tends to overcome the stone at 1 m from the support axis.

Answer

Data	Request	Formula
BM = 1 m BR = 20 cm FR = 2.10^3 N	FM = ? in N	FMBM = FRBR
Reasoning		
FM x 1 = 2000 x 0.02; FM = 400 N		

2. A 200 N stone is lifted with a 1.10 m bar. Calculate the lever arm of the 20 N driving force. **Answer: 1 m.**

3. A load of 1200 N is lifted with a wheelbarrow whose wheel axis is located 40 cm from the point of application of the load and the driving force is exerted 120 cm from the wheel axis. Calculate the bearing force on the wheel axle. **Answer: 800 N.**

3. Scourge

The flail is an inter-lever used in the balance that compares the weights of the body mass and the mass marked on it. The flails are in horizontal balance under the conditions such as:

$F1l1 = P2l2$

where F : component of the forces in N or in kgf

l: arm of the flail in meters.

N.B.: The moments of the forces with respect to the support axis are equal.

Application exercises

1. A metal fragment M placed in pan A of a balance is balanced by a mass of 7200 g in pan B. The same fragment placed in pan B is balanced by a mass of 7180 g in pan A. Calculate the arm ratio of the beam.

Answer

Data	Request	Formula
$M_1 = M$ $m_2 = 7200g$ $m_1 = 7180$ g $m_2 = M$	$\frac{l_1}{l_2} = ?$	$P_{11} l_1 = P2l2$
Reasoning		
Condition 1; $Mgl1 = 7200gl_2$ $\frac{l_1}{l_2} = \frac{7200}{M}$ (1) Condition 2; $7180gl_1 = Mgl_2$ $\frac{l_1}{l_2} = \frac{M}{7180}$ (2) (1) = (2) $\frac{7200}{M} \frac{M}{7180}$; $M^2 = 7200 \times 7180$ $M^2 = 51696000$ M = 7189.9Kg M = 7190Kg $\frac{l_1}{l_2} = \frac{7200}{7190} = 1$		

2. A mass placed in pan A of a balance is balanced by a mass of 4 g in pan B. The same mass placed in pan B is balanced by a 9 g mass in pan A. Calculate the arm ratio of the beam. **Answer: 0.66 or 1.**

Ref : J. Jodogne ; Physique 1, Mécanique ; A De Boeck 1970, p106.

4. Beam

The beam is a rigid bar resting on a support. The beams are in equilibrium under conditions such as:

$$\mathbf{F_1 AC = F_2 BC}$$

where

F_1 and F_2 : components of forces in N,

AC and BC: segment (distance) in meters,

A, B and C: points of application.

N.B.: Two parallel forces of the same direction and the same sense have a resultant :

$$\mathbf{F = F_1 + F_2}$$

Two parallel forces of the same direction but opposite sense have a resultant :

$$\mathbf{F = F_2 - F_1}$$

Ref : E. PIRA; Energie, Kinshasa, 1973, p 80.

Application exercises

1. Two parallel forces of 1000 N and 800 N act on each other at the ends of a 1.8 m long beam.
   a) Calculate the resultant of these parallel forces having the same direction and the same sense.
   b) Calculate the distance at which the resultant of parallel forces having the same direction and sense acts.

Answer

Data	Request	Formula
AB = 1.8 m F_1 = 1000 N F_2 = 800 N	a) F = ? in N b) BC = ? in m	a) $F = F_1 + F_2$ b) $F_1 AC = F_2 BC$

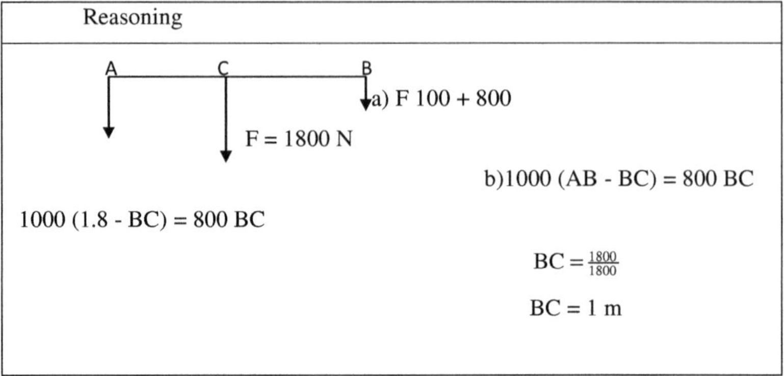

2. Two horses coupled to a carriage by means of a 1.8 m spreader bar are pulling one with a force of 100 kgf and the other with a force of 80 kgf. Calculate the distance at which the resultant of these two forces having the same direction and the same sense applies. **Answer: 1 m**

3. Two parallel forces of 6 kgf and 10 kgf act on each other at points of application which are 80 cm apart. Calculate the distance of the point of application of the resultant of these parallel forces having the same direction. **Answer: 50 cm.**

4. The ends of a 5 m beam rest on two walls. At 2 m from the first wall, a 50 kgf box is suspended. Calculate the force on each wall, neglecting the weight of the beam. **Answer: 30 and 20 kgf.**

5. A force of 9 kgf acts at 20 cm from the end of a 36 cm rod. Calculate the force decomposed into two parallel components of the same direction and the same sense having their points of application at the ends of the rod. **Answer: 4 and 5 kgf**.

6. A 5 m beam is embedded in a wall by one of its ends. At 3 m from the wall, it rests on a support. At the free end, a weight of 45 kgf is suspended. Calculate

a) The force exerted on the support if we neglect the weight of the beam.

Answer: 75 kgf.

b) Calculate the force exerted on the wall. **Answer: 30 kgf.**

6. A 2.5 m bar has one end fixed in a wall and the other end hangs a 45 kgp weight. At 1.5 m from the wall, the bar rests on a column. Calculate:

a) The component force that acts on the wall. **Answer: 75 kgf**.

b) The component force that acts on the column. **Answer: 30 kgf.**

7. Two parallel forces of 3 kgf and 7 kgf are located 40 cm apart. Calculate the distance of the point of application of the resultant of these forces having opposite directions. **Answer: 30 cm.**

8. Two parallel forces of 1400 N and 2600 N are located 90 cm apart. Calculate the distance of the point of application of the resultant of these parallel forces of opposite directions. **Answers: 1.5 cm.**

5. Key or key

The key is a metal used to tighten the nut and to open a beer or a door. The keys exert two parallel forces which constitute a driving torque of one moment:

$$\mu = Fd \sin \alpha$$

where μ : moment of a motor in Kgfm or in Nm ;

F : force in N or kgf,

d : distance in meters,

α : angle in degrees.

N.B.: The moment of a pair of forces corresponds to the work of a force.

Application exercises

1. A 25 cm wrench is used to tighten a nut with a force of 5 kgf. The directions of the force form an angle of 60°. Calculate the moment of the applied force with respect to the axis.

Answer

Data	Request	Formula
$F = 5kgf$ $d = 25cm$ $\alpha = 60°$	μ = ? in Kgf.m	$\mu = Fd \sin\alpha$
Reasoning		
$\mu = 5 \times 25.10^{-2} \sin 60°$ $\mu = 625 \times 10^{-3} \sqrt{3}$ $\mu = 625 \times 10^{-3} \times 1.73$ $\mu = 1081$ kgf.m		

2. Two parallel forces of 0.3 kgf act on a body, have their points of application in A and B at the distance of 0.4 m. Calculate the moment of two forces that each form an angle of 30° with a line AB. **Answer: 0.06 kgf. m.**

3. A 1 dm corkscrew is used to open a beer bottle with a force of 0.6 kgf. The directions of the force form an angle of 45°. Calculate the moment of the applied force relative to the cork. **Answer: 423 x 10 $^{-4}$ kgf. m.**

Ref : E. Pira, Energie; Kinshasa, 1973, p77.

6. Cable

The rope is a metal rope made of twisted wires. The ropes are stretched by the action of two extraordinary forces having a resultant:

$\mathbf{F = F_1 + F_2}$

where F : resultant of the forces in Newtons ;

F_1 and F_2 : components of forces in N.

Application exercises

1. Two men pull a beam with the same rope, one exerts a force of 50 kgf and the other exerts a force of 40 kgf. Calculate the resultant of these forces.

Answer

Data	Request	Formula
$F_1 = 50$ kgf $F_2 = 40$ kgf	F = ? in kgf	$F = F_1 + F_2$
Reasoning		
F = (50 + 40) kgf F = 90 kgf.		

2. A crane lifts a 100 N load with a force of 180 N. Calculate the resultant of these components. **Answer: 80 N.**

3. With the help of a cable, a force of 60 kgf is exerted on a leaning shaft which opposes with a force of 36 kgf. Calculate the resultant of these forces. **Answer: 24 kgf.**

4. Using a cable, a force of 80 kgf is exerted to pull a leaning coconut tree on the roof of the hospital with a force of 90 kgf. Calculate the resultant of these components. **Answer: 10 kgf.**

5. Two students each hold the same force of 6 kgf in opposite directions. Calculate the resultant of these forces. **Answer: 0 kgf.**

6. Two forces of 4 kgf and 7 kgf are exerted by two men pulling a load on the same string. Calculate the resultant of these forces exerted in opposite directions. **Answer: 3 kgf.**

Ref: E. Pira; energy; Kinshasa, 1973, p67.

7. Fixed pulley

The fixed pulley is a fixed disc with a groove through which a rope used to lift a load passes. The fixed pulley is supported by a clevis with a hook that is subjected to a supporting force:

$$\mathbf{F_A = F_R + F_M + P_P = 2F + P_p}$$

where F_A: support force in Newtons,

Pp : weight of the pulley in Newtons.

N.B.: The equilibrium conditions of the fixed pulley are achieved by equality of the driving force and the resisting force applied to the cable (rope) goods.

Application exercises

1. A 100 kg bag is lifted with a 10 kg fixed pulley. Calculate the bearing force on the fixed pulley.

Answer

Data	Request	Formula
Pp = 10 kg F = 100 kg	F_A = ? in N	$F_A = 2F + Pp$
Reasoning		
$F_A = 2 \times 100\ kg \times 9.8\ m.s^{-2} + 10 \times 9.8\ m.s^{-2}$ $F_A = 2058\ N$		

2. A 180 kg crate is lifted by means of an 18 kg fixed pulley. Calculate the bearing force on the fixed pulley. **Answer: 3704.4 N**

3. A fixed pulley of 80 kg is in equilibrium when a driving force is exerted on a load of 200 kg box. Calculate the bearing force of the fixed pulley.

Answer: 4704 N.

4. A fixed pulley of 98 N is balanced by a load of 490 N. Calculate the bearing force of the fixed pulley. **Answer: 1078 N.**

5. A 200 N stone is lifted with a 20 N fixed pulley. Calculate the bearing force of the fixed pulley. **Answer: 420 N.**

Ref : J. Jodogne ; physique 3 ième ; A De Boeck 1970, p 103.

8. Movable pulley

Objectives

Knowledge: define the pulley

Know-how: solve the exercises on the pulley

The movable pulley is a movable disc that slides on a cable (rope) used to lift a load. The mobile pulleys allow to lift a load by a driving force:

$$F = \frac{P}{2}$$

where F : driving force,

P : weight of the load.

N.B. : A motor work is a driving force. A resisting work a resisting force.

Application exercises

1. A mobile pulley lifts a load of 50 kgf to a height of 3m. Calculate the work of a resisting force and a driving force.

Answer

Data	Request	Formula
F = 50 kgf h = 3 m	T = ? in J	T = Fd
Reasoning		
a) $F_R = 50$ kgf x 9.81 m.s $^{-2}$ T = 50 x 9.81 x 3 T = 1472 J b) $F_M = \frac{50}{2}$ x 9,81 and h = 2 x 3 $T = \frac{50}{2}$ x 9,81 x 6 T = 1472 J		

2. A mobile pulley lifts a 100 kg bag to a height of 5 m.

a. Calculate the work of a driving force. **Answer: 490 J**

b. Calculate the work of a resistive force. **Answer: 4905 J**

3. A 5N mobile pulley lifts a 100N load to the height of 2m.

a. Calculate the work of a driving force. **Answer: 210 J**

b. Calculate the work of a resistive force. **Answer: 210 J**

4. A mobile pulley lifts a 2 ton load. Calculate the driving force required to lift the load. **Answer: 19620 N**

Ref : J. Jodogne ; physique 3 ^{ième} ; A De Boeck 1970, p 112.

9. String

Objectives

Knowledge: define the string

Know-how: calculate the elongation of the cable

The string is a wire attached to a support that moves along a vertical line. The strings wound twice around the drums (rotating discs) of an elevator that changes the direction of the force are comparable to springs subjected to the action of a tensile force, whose formula is :

$F = E.S.\Delta l/l$

Where Δl: elongation in meters,

F : force in Newtons,

l : length in meters,

S : section in square meters,

E : Young's modulus in N. m-2.

N.B.: the tensile force is exerted on a spring which undergoes an elastic deformation.

Ref : E.PIRA ; Energie, mécanique ; Kinshasa, 1973, p 80.

Application exercises

1. A wire rope of section 3 cm and length 10 m supports an elevator of 4 tons. Calculate the elongation of the rope in the case where the Young's modulus is 22 x103 kgf.mm-2.

Answer

Data	Request	Formula
$S = 3\ cm^2$ $L = 10\ m$ $F = 4\ tons$	$\Delta l = ?$ in mm	$\Delta l = \frac{FL}{ES}$
Reasoning		
$\Delta l = \frac{4.10^2 \times 10^4}{22.10^3 \times 300}$ $\Delta l = 6\ mm$		

2. A wire of 1 mm ^{2} and 1 m long is subjected to the action of a weight of 30 kgp. Calculate the elongation of the wire under the conditions where E = 20 x ^{103} kgf.mm $^{-2}$

Answer: 1.5 mm.

3. A wire of 10 m and 5 mm ^{2} is extended to 3 m under a tensile force of 24 kgf. Calculate the Young's modulus in SI. **Answer : 15,68 x 10 ^{8} N.m $^{-2}$**

4. A 33 m, 0.8 cm2 wire rope used for an elevator that supports people extends 15 cm. Calculate the weight of the elevator. **Answer: 8 x 10 $^{-3}$ kgp.**

10. Coil spring

Objectives

Knowledge: What is a coil spring?

Know-how: calculate the period of oscillations of a spring

The coil spring is a wire in the form of coils which is deformed by a load. Coil springs follow a period of oscillations given by:

$$T = 2\pi\sqrt{\frac{m}{k}}$$

where T : period of the oscillations in seconds,

m: mass of the body in Kilograms,

k : spring stiffness in N.m-1.

Application exercises

1. A mass of 0.02 kg is attached to a spring of stiffness 2.8 N.m $^{-1}$. Calculate the period of oscillation of a spring.

Answer

Data	Request	Formula
m = 0,02 Kg k = 2.8 N.m $^{-1}$	T = ? in s	$T = 2\Pi\sqrt{\frac{m}{K}}$
Reasoning		
$T = 6{,}28\sqrt{\frac{0{,}02}{2{,}8}}$ $T = 6{,}28\sqrt{\frac{2}{280}}$ T = 0.53 s		

2. A coil spring extends 35 mm when a 10 gp weight is suspended from it. Calculate the stiffness of the spring. **Answer: 2.8 N.m $^{-1}$.**

3. An elevator ascends with a load of 100 kg. Calculate the apparent weight of the load carried by the elevator. **Answer: 1000 N**.

4. A 25 kg child is supported by a scale ...person, then travels upwards.

During the swing, the needle reaches a scale that corresponds to 27 kg. Calculate the acceleration of the child's movement. **Answer: 1.08 m.s $^{-2}$**

Ref : E. PIRA; Energie, mécanique; Kinshasa, 1973, p 116.

CHAPTER IV: DYNAMICS

IV.1 DYNAMIC EFFECTS

1. Mobile

Objectives

Knowledge: give the dynamic effects of the force

Know-how: calculate the resultant of forces

The mobile is a body animated by a movement or any body that moves or in motion. Motives are subject to the action of concurrent forces having a resultant:

F = $\sqrt{F_1^2 + F_2^2 + 2F_1F_2\cos\alpha}$

Where F : resultant of the forces in N or kgf ;

F_1 and F_2 : components of the forces in N or in kgf.

α : angle

N.B.: The perpendicular forces correspond to the angle $\alpha = 90°$ and cos 90° = 0.

Ref : E. PIRA; energy; Kinshasa, 1973, p 78.

Application exercises

1. Two concurrent forces of 6 Kgf and 10 Kgf acting on a body form an angle of 60°. Calculate the resultant of these forces.

Answer

Data	Request	Formula
F_1 6 kgf F_2 10 kgf $\alpha = 60°$	F = ? in Kgf	$F^2 = F_1^2 + F_2^2 + 2F_1F_2\cos\alpha$
Reasoning		
$F^2 = 36$ kgf + 100 kgf + 120 kgf cos60 $F = \sqrt{196}$ kgf $= 14$kgf		

2. Two concurrent forces of 60 newtons act on each other perpendicularly on a body. Calculate the resultant of these forces.

Answer: 84.8 N.

3. A mass of 0.15 grams is subjected to the action of two perpendicular forces of respective intensities of 3 and 4 newtons. Calculate the acceleration of the mass. **Answer : 33 x ^{103} m.s $^{-2}$.**

4. Two perpendicular forces of 60 N and 40 N act on a body. Calculate the resultant of these forces.

Answer: 72.1 N.

5. A force of 8 kgf is decomposed into two concurrent forces F_1 and F2 of 6 kgf. The directions of force F_1 and force F2 form an angle of 90°. Calculate the force F_1. **Answer: 5.1 kgf.**

2. Inclined plane

The inclined plane is a slope that facilitates the descent of a body. The inclined plane allows a body to descend animated by a driving force:

$$\mathbf{F} = \mathbf{P}\frac{h}{l} = \mathbf{P} \sin \alpha$$

where F : driving force in Newtons,

P : body weight in Newtons,

h: height in meters,

l : length in meters,

$\sin\alpha$: difference in level.

Application exercises

1. A wagon goes down on an inclined plane of length 400 m and height 16 m. Calculate the time taken by the wagon on the inclined plane.

Answer

Data	Request	Formula
l = 400 m h = 16 m	t = ? in s	$e = \frac{1}{2} at^2$

Reasoning
$F = mg\frac{h}{l} = ma$ $a = 10\ m.s^{-2}\ \frac{16}{400}\frac{m}{m} = 0{,}4\ m.s^{-2}$ $e = \frac{1}{2} a\ t^2 \Rightarrow t^2 = \frac{2e}{a} = \frac{2 \times 400}{0{,}4}$ $t = 48{,}1\ s$

2. A 20 ton vehicle descends on an inclined plane with a 1% difference in level. Calculate the force exerted on the vehicle.

Answer: 1960 N.

3. An empty 60-kilogram trailer rolls on a 30° inclined plane. Calculate the force exerted on the empty trailer. **Answer: 294 newtons.**

4. A 1500 kilogram wagon descends on an inclined plane of length 0.5 m and height 5 cm. Calculate the force exerted on the car.

Answer: 588 x ^{103} N.

5. A cart descends a slope of 80 meters with a 50% difference in level. Calculate the time taken by the cart. **Answer: 5.7 seconds.**

Ref : J. Jodogne ; physique 3ième ; A De Boeck 1970, p 60.

3. Engine

Objectives

Knowledge: define an engine

Know-how: calculate the power developed on a bucket of water

The motor is a body that exerts a force to produce work.

The motors develop in operation a power :

$$\mathbf{P} = \frac{T}{t}$$

where P : power in watts,

T : work in joules,

t : time in seconds.

N.B.: 1 Horsepower (HP) = 736 watts.

Ref : J. Jodogne ; physique 3 ième ; A De Boeck 1970, p 115.

Application exercises

1. A mass is suspended from the mount with a 0.2 w bulb in one hour. Calculate the mass suspended from the mount at the height of 2 m with the bulb.

Response:

Data	Request	Formula
P = 0.2 w t = 1h = 36000 sec	m = ? in kg	$P = \frac{T}{t}$
Resolution		
$P = \frac{mgh}{t}$; $m = \frac{P\,t}{g\,h}$ $m = \frac{0,2 \text{ x } 3600}{10 \text{ x } 2}$ m = 36 kg		

2. A 15 kg clock motor rises every 8 days from a height of 2 meters. Calculate the average power required to operate the shade. **Answer: 5.78 x 10 $^{-7}$ HP**.
3. A 5 t truck moves over a distance of 200m in seconds. Calculate the power developed by the engine. Answer: **490.5 kilowatts.**
4. A crane that lifts a load of 450 kg to a height of 5 m develops a power of 18 HP. Calculate the time after which the load is lifted by the crane.

Answer: 1.66 s.

5. A bucket of water of 80 N is lifted from a 15 m deep well in 1 minute. Calculate the power developed on the bucket of water. **Answer: 20 w.**
6. A motor develops a power of 1 kilowatt in one hour. Calculate the work done by the motor. Answer: **3.67 kgfm.**

4. Chain

The chain is a body that fits into the teeth of a sprocket and a pinion. The chains that transmit the rotational motion of the sprocket to the sprocket produce a development of a bicycle:

$$\textbf{dev} = \pi D\frac{N}{N'}$$

where dev: development of a bicycle in meters,

D : diameter of the gear wheel,

N : number of teeth of the gear wheel,

N' : number of teeth of a pinion.

N.B.: The ratio of the chain transmission speeds corresponds to the formula :

$$\frac{N}{N'} = \frac{d}{D}$$

Application exercises

1. A chain that transmits rotational motion from the 46-tooth sprocket to the 18-tooth sprocket rotates at a rate of one revolution of the crankset. Calculate the ratio of chain drive speeds.

Response:

Data	Request	Formula
N = 46 N'= 18	$\frac{N}{N'} = ?$	$\frac{N}{N'} = \frac{46}{18}$
Reasoning		
$\frac{N}{N'} = \frac{23}{9}$; 1 turn of the crankset corresponds to $\frac{46}{18}$ turns of the sprocket and $\frac{23}{9}$ turns of the pinion		

2. A chain that transmits a rotational motion from the 46-tooth sprocket to the 18-tooth sprocket rotates at a rate of one revolution of the crankset. Calculate the development of the bicycle that corresponds to the rotation of the sprocket by 0.7m. **Answer: 5.35 m.**
3. A belt transmits a rotational motion from the 20 cm diameter pulley to the 8 cm diameter pulley. Calculate the ratio of the speeds transmitted by the belt. **Answer:** $\frac{4}{10}$

Ref : J. Possoz ; Technologie 2 ième année secondaire, Kinshasa, 1989, p 45

5. Wheel

The wheel is a body on which a moving vehicle moves. The wheels are animated by a rotational speed that corresponds to the linear speed:

$$\mathbf{v = 2\pi\, rf = \varpi\, r}$$

where

v : linear velocity in m. S^{-1},

r : radius of the wheel in meters,

f : frequency of the movement in revolutions per second or Hertz,

ϖ : angular velocity in rad. S^{-1}.

N.B.: the frequency of the motion is the inverse of the period of the motion evaluated in seconds; π (pi) = 3.14.

Ref : J. Possoz ; Technologie 2 ième année secondaire ; Kinshasa, 1989, p 43.

Application exercises

1. A wheel describes a circumference of diameter 6 dm at a rate of 15 revolutions per second around a fixed axis. Calculate the rotational speed which corresponds to the linear speed.

Response:

Data	Request	Formula
v = 15 trs. s^{-1} D = 6 dm	v = ? in m. s	$v = 2\pi$ v r
Reasoning		
$v = 6{,}28 \times 15 \times \frac{0{,}6}{2}$ v = 6.28 x 15 x 0.3 v = 28.26m/s		

2. A wheel of diameter 4 dm rotates at a rate of 54 revolutions per minute around a fixed axis. Calculate the speed of rotation of the wheel. **Answer: 113.04 m. s^{-1}.**

3. A wheel with a diameter of 60 cm rotates at a rate of 120 revolutions per minute. Calculate the linear speed and the angular speed of the wheel.

Answer: a) 3.77 m. s^{-1} b) 12.56 rad. s^{-1}.

4. The diameters of a gear and a pinion are 80 cm, 18 cm and 6 cm respectively. Calculate the speed of the cylinder when the crankset makes 60 revolutions per minute. **Answer: 7.54 m. s^{-1} (27 kilometers per hour).**

6. Simple pendulum

The simple pendulum is an oscillator that moves from and to the equilibrium position by a force of gravity. The simple pendulum follows the period of oscillations :

$$\mathbf{T = 2}\ \pi\sqrt{\frac{l}{g}}$$

where T : period of the oscillations in seconds,

l: length of the wire in meters,

g : acceleration of gravity,

π : pi.

N.B.: pi is equivalent to the value of 3.14.

The period of the oscillations corresponds to the inverse of the frequency (f), evaluated in Hertz (Hz):

$$\mathbf{T} = \frac{1}{f}$$

Application exercises

1. A simple pendulum with a wire length of 2.45 m produces oscillations in a medium where g = 9.8m. S^{-2}. Calculate the frequency of the motion of a simple pendulum.

Answer

Data	Request	Formula
l = 2,45 m g = 9.8 m. s^{-2}	N = ? in Hz	$T=\frac{1}{N}$
Reasoning		
$T=2\pi\sqrt{\frac{l}{g}}=\frac{1}{N}; T=6{,}28\sqrt{\frac{2{,}45}{9{,}8}}$ $T=6{,}28\sqrt{0{,}25}$ T = 6.28 x 0.5 = 314.10-2 s T = 3.14 s $N=\frac{1}{T}$ $N=\frac{1}{3{,}14}=0{,}31\,Hz$		

2. A simple pendulum beats one second at a place where g = 9.812 m. s $^{-2}$. Calculate the length of the wire of a simple pendulum. **Answer: 0.09 m.**

3. A simple pendulum beats every half second. Calculate the frequency of a pendulum's motion. **Answer: 0.5 Hz.**

4. A simple pendulum with a wire length of 2.45 m swings in a medium where g = 9.8 m. s $^{-2}$. Calculate the period of the oscillations of a pendulum. **Answer: 3.14 s.**

5. A simple pendulum of 150 g suspended from a wire of length 4 m is moved away from its equilibrium position by an angle of 60°. Calculate the work done by the simple pendulum moving away from the equilibrium position. **Answer: 3 J.**

Ref : E. Pira; energy; Kinshasa, 1973, p113

7. Satellite

The satellite is a planet that produces a revolutionary motion around a star on the elliptical orbit. Satellites gravitate on the elliptical orbit by a force of gravitational interaction defined by the universal law of gravitation:

$$F = 6{,}67.10^{-11} \frac{m_1 m_2}{r^2}$$

where : force in Newtons,

m1 and m2 : masses in Kilograms,

r : orbital radius in meters.

N.B.: The mass of the earth m = 6 x 10 ^{24} kilograms,

The terrestrial radius rt = 6400 kilometers,

The orbital radius r = rt + h because h is the altitude.

Application exercises

1. Two spheres of 300 kg are located center to center at a distance of 5 m. Calculate the universal force of attraction exerted on the two spheres.

Response:

Data	Request	Formula
m = 300 kg d = 5 m	F = ? in N	$F = 6{,}67.10^{-11} \frac{m_1 m_2}{d^2}$

Reasoning
$F = 6{,}67 \times 10^{-11} \frac{(300kg)^2}{25m}$ F = 239 x 10-9 N

2. Two objects of 10 tons are placed center to center at 20 m. Calculate the universal force of attraction exerted on the two objects. **Answer: 1, 7 x 10^{-7} N.**

3. The earth exerts a universal force of attraction on a sphere of 10^{4} kg. Calculate the force of universal attraction exerted by the earth on the sphere. **Answer: 98 x 10^{4} N.**

4. A sphere of 10^{4} kg is placed on the human surface of 7 x 1022 kg and radius 1740 km. Calculate the universal force of attraction exerted by the moon on the sphere. **Answer: 16 x 10^{6} N.**

5. A satellite orbits at an altitude of 600 km. Calculate the acceleration of gravity of a satellite orbiting the earth.

Answer: 8.2 m. s^{-2}.

6. A 900 kg satellite orbits at an altitude of 600 km around the earth. Calculate the weight of the satellite. **Answer: 8 x 10^{2} N.**

Ref : A. Bribosia ; physique 5ième, option complémentaire ; A De Boeck 1992, p 59.

8. Rochet

The ratchet is a solid that falls each time in the hollows of a ratchet. The ratchet moves from one tooth to the other of a ratchet when the drum of a winch is turned around an axis by a crank under driving force:

$$F = p\frac{r}{l}$$

where F : driving force in Newtons,

r: radius of the drum,

l: crank arm,

p: weight.

N.B.: With n turns of the crank, the load increases to the height :

$h = 2\pi rn$

The latch is used to immobilize a load and prevent the load from moving out.

Application exercises

1. A winch with a cylinder diameter of 75 mm and a crank arm of 0.5 m lifts a load of 200 kg.

   a) Calculate the force exerted on the crank to lift a load.
   b) Calculate the height of the load lifted with 4 turns of the crank.

Answer.

Data	Request	Formula
m = 200 kg r = 75 mm l = 0.5 m	a) F ? in N b) h = ? in m	a) $F = P\frac{r}{l}$ b) $h = 2\pi r\, n$
Reasoning		
a) $F = 200 \times 9{,}81 \frac{0{,}075}{0{,}5}$ F = 294.5 N b) h = 2 π x 0.075 x 4 h = 1.885 m		

2. A winch with a drum diameter of 20 cm lifts a load of 3000 N. Calculate the arm of the crank that rotates with a force of 300 N. **Answer: 1 meter**.

3. A winch with a cylinder radius of 10 cm and a crank arm of 50 cm lifts a load of 200 kg. Calculate the work of the resisting force and the work of the driving force for one revolution of the crank. **Answer: 1232.136 J and 12321.16 J.**

Ref : J. Jodogne ; Physique 1, mécanique ; A De Boeck 1970, p 113.

9. Rotating arrow

The rotating boom is an upper part that gives the crane its characteristic shape. The rotating jib is one of the essential parts of the crane which will be in balance in the conditions such as:

P x a = P' x b

where P : weight of the crane,

P' : weight of the load,

a and b: lever arms.

N.B.: the crane will be stable when $P > P'\frac{b}{a}$; b>a.

Application exercises

1. A crane with lever arms of 4 m and 9 m respectively lifts a load of 20 tons. Calculate the weight of the crane.

Response:

Data	Request	Formula
$b = 9$ m $a = 4$ m $P' = 20$ t	$P = ?$ in Kgp	**P x a = P' x b**
Reasoning		
$P = P'\frac{b}{a}$ $P = 2 \times 10^3 x \frac{9m}{4m}$ $P = 45 \times 10^{3}$ kgp		

2. A crane with lever arms of 3 and 6 m respectively, weighs 40 tons. Calculate the weight of the load lifted by the crane. **Answer: 20 x 10 ^{3} kgp.**

3. A 20 ton crane lifts a 2 ton load. Knowing that the lever arms are 3 and 8 m. Is the crane stable? Why is it stable? **Answer: Yes, because 20>5.3.**

Ref : J. Possoz ; Technologie 2 ième année secondaire ; Kinshasa, 1989, p 17.

IV.2 FLUID STATICS

1. Pressure

Knowledge: give the dynamic effects of the force

Know-how: calculate the resultant of forces

Pressure is a thrust caused by a solid pushing a fluid. Pressure is transmitted in all directions and the useful ones actuate certain tools, notably: the inner tube, the brakes of large trucks and the train.

Atmospheric pressures exert a pressing force on any body encountered in space (in a vacuum).

The pressure caused by boiling water raises the boiling temperature and the pressure caused by the solidification of water lowers the boiling temperature.

The pressures of the water flowing from the tap depend on the height of the water tower and a cistern. The water vapor pressure of 200 kgf. cm $^{-2}$ corresponds to the critical temperature of water of 365 ° C and that of 16 kgf. cm $^{-2}$ called stamp, corresponds to the temperature of 200 ° C. The saturation pressure corresponds to the pressure of saturated water vapor in air. Pressures are evaluated in the MKSA system of units in Newtons per square meter (N. m $^{-2}$) called Pascal ($_{Pa}$), in atmosphere (atm), in bar (b) and in dyne per square centimeter (dyn. cm $^{-2}$). They are also evaluated in the system of units MKf. S in Kgf. m $^{-2}$ and in gf. cm $^{-2}$.

N.B.: Under normal (standard) temperature conditions, the atmospheric pressure is 1 atmosphere (atm).

1 N. m $^{-2}$ = 1 $_{Pa}$ = 100 dyn. cm $^{-2}$ = 10 $^{-5}$ bar

1 kgf. m $^{-2}$ = 9,81 N. m $^{-2}$ = 9,81 P $_a$

1 atm = 10 ^{5} P $_a$ = 1033 gf. cm $^{-2}$ = 1 bar

1 gf. cm $^{-2}$ = 981 dyn. cm $^{-2}$ = 981 x 10 $^{-6}$ bar = 0,981 mb

Ref : J.JODOGNE ; Physique 1, Mécanique ; A De Boeck 1970, p.153.

2. Piston

The piston is a solid that produces a reciprocating motion in the cylinder. The pistons develop a pressure on the fluids calculated by the relation :

$P = \frac{F}{S}$

where P : pressure in N. m $^{-2}$ or in kgf. cm $^{-2}$

F : force in N or kgf

s : surface (section) in square meters (m ^{2}).

N.B.: The surface, $S = \pi r^2$

Application exercises

1. In an automobile elevator, the diameter of the piston subjected to the force of 1250 kgf is 20 cm. Calculate the pressure of the compressed air in the elevator.

Response:

Data	Request	Formula
F = 1250 kgf D = 20 cm	P = ? in Kgf.cm $^{-2}$	$P = \frac{F}{S}$
Reasoning		
$P = \frac{1250}{3{,}14 \times 10^2}$ P = 4 kgf.cm-2		

2. A book leaning on a table of 330 cm ^{2} exerts a rectangular force of 500 grams force. Calculate the pressure developed by the book.

Answer: 149 N. m $^{-2}$

3. A statue of 20 tons is resting on a surface of the ground of 6 cm ^{2}. Calculate the pressure developed by the statue.

Answer: 32.7 x 10 ^{4} N. m $^{-2}$

4. In the boiler of a steam engine, the pressure which discharges the steam is 7.5 kgf. cm $^{-2}$. Calculate the force exerted on the wall of 1 m ^{2} of section.

Answer: 736 x 10 ^{4} N

3. Fluid

The fluid is a body considered as a gas or a liquid. The fluids transmit completely the pressures in the communicating vessels with a volumic weight:

$\varpi = \frac{P}{V}$

where $\bar{W}$: weight by volume in N.m $^{-2}$ or in kgf. m $^{-3}$,

P: weight in N or kgp (kg'),

V : volume in m ^{3}.

N.B.: the volume weight is measured by a hydrometer which exists as a float of particular shape.

Ref : J. Jodogne ; Physique 1, mécanique ; A de Boeck 1970, p195.

Application exercises

1. A mercury that occupies a volume of 80 cm ^{3} weighs 7200 gp. Calculate the weight by volume of mercury.

Answer

Data	Request	Formula
P = 7200 gp V = 80 cm ^{3}	$\bar{W}$ = ? in N. m $^{-3}$	$\bar{W} = \frac{P}{V}$
Reasoning		
$\bar{W} = \frac{7200 \times 9{,}81}{80.10^{-6}}$ $\bar{W}$ = 88 290 N. m $^{-3}$		

2. An alcohol that occupies a volume of 70 cm ^{3} weighs 3000 gp. Calculate the weight by volume of alcohol. **Answer : 42,1 x 10 ^{5} N. m $^{-3}$**

3. A mercury occupies 50 cm3 with a weight of 8200 gp. Calculate the weight by volume of mercury. **Answer : 13,3 x 10 ^{5} N. m $^{-3}$**

4. Air that occupies 100 cm3 weighs 8200 gp. Calculate the weight by volume of air. **Answer: 80.4 x 10 ^{5} N. m $^{-3}$**

4. Liquid

The liquid is a fluid in which the molecules exert a mutual attraction between them that produces a weak cohesion of medium distance.

Liquids rise in volume with a density :

$d = \frac{\rho}{\rho_0}$

where d: density,

ρ density of the considered liquid,

ρ_0: density of the reference liquid.

N.B.: The density of the body is calculated by : $\rho = \frac{m}{V}$

The density of water is 1g. m $^{-3}$.

The density of mercury is 13.6 g. m $^{-3}$.

Ref : J. Jodogne ; Physique 3, chaleur ; A De Boeck 1969, p17.

Application exercises

1. A 680 g mercury occupies the same volume with 50 g distilled water. Calculate the density of mercury under standard conditions of temperature and pressure.

Answer

Data	Request	Formula
$M_{Hg} = 680$ g $M_e = 50$ g	d = ?	$d = \frac{\rho}{\rho_0}$
Reasoning		
$\rho = \frac{m}{v}$; $d = \frac{m_H/V}{m_e/V}$; $V_{Hg} = V_e$ $d = \frac{680}{50} = 13{,}6$		

2. A mercury of 680 g occupies 40cm3 under normal conditions of pressure and temperature. Calculate the density of mercury. **Answer: 12 x 10 $^{-4}$**

3. A water of 988 g occupies 500 cm3 under normal conditions of pressure and temperature. Calculate the density of the water. **Answer: 1.97**

4. A block of copper with a density of 8.9 g. cm $^{-3}$ is immersed in 0.1 liter of water. Calculate the mass of the copper block. **Answer: 0.89 kg**

5. A silica cube is immersed in water. To restore equilibrium, 15 g of the side of the silica cube is added. Calculate the volume of the cube. **Answer: 125 cm ^{3}**

6. A water of 1 kg solidifies to obtain an ice of 1090 cm ^{3}. Calculate the density of the ice. **Answer: 91.743 kg. m $^{-3}$**

5. Mercury

Mercury is a liquid contained in the orifice of a capillary tube.

Mercury rises to the height of the column of a capillary tube under pressure:

$P = \bar{W} h$

where p : pressure of mercury in atmospheres or in kgf. m-3

h : height of the column in meters.

N.B.: The height of the mercury column corresponds to the pressure of 76 cm of mercury.

The weight by volume: $\bar{W} = p\ g$

The density of mercury is: $\bar{W} = 13.6$ gp. cm $^{-3}$

Ref : J. Jodogne ; Physique 1, mécanique ; A De Boeck 1970, p82.

Mercury Exercises

1. In the MKSA system, the density of mercury is 9.81 x 13590 N. m $^{-3}$ and the height of the mercury column is 0.76 m. Calculate the atmospheric pressure exerted by the atmosphere.

Answer

Data	Request	Formula
$\bar{W}_{Hg} = 9.81 \times 13590$ N. m $^{-3}$ h = 0.76 m	P = ? in atm	$P = \bar{W}\ h$
Reasoning		
P = 9.81 x 13,590 x 0.76 P = 10 1300 Pa = 10 ^{5} Pa P = 1 atm		

2. In the CGS system, the density of mercury is 981 x 13.59 dyn. cm^{-3} and the height of the mercury column is 76 cm. Calculate the atmospheric pressure exerted by the atmosphere. **Answer: 0 atm.**

3. In the MKS system, the density of mercury is 13590 kgf. m^{-3} and the height of the mercury column is 0.76 cm. Calculate the pressure exerted by the atmosphere. **Answer: 10330 Kgf. m^{-2}.**

4. In the international system of units, the height of the mercury column is 75 cm and the density of mercury is 13600 Kg. m^{-3}.

Calculate the atmospheric pressure exerted by the atmosphere.

Answer: 10^{5} N. M^{-1}, or 1 atmosphere.

5. A water with a density of 1 gf. cm-3 rises to the height of the column of 25 cm of mercury. Calculate the pressure exerted by the water. **Answer: 25 gf. cm^{-2}.**

6. An iodinated alcohol with a density of 0.8 gf. cm^{-3} rises to the height of the mercury column of 25 cm. Calculate the pressure exerted by the iodinated alcohol.

Answer: 20gf. cm^{-2}.

7. A water with a density of 1.026 gf. cm^{-3} has a depth of 300 meters. Calculate the pressure of the sea water that corresponds to the height of the mercury column. **Answer: 3078 gf. cm^{-2}.**

8. A cylindrical vase contains mercury at a height of 3 m and water at a height of 20 cm.

Calculate the force exerted on the bottom of the vase with a surface area of 80 cm^{2}. **Answer: 4864 gf.**

9. A test tube contains mercury on the height of 10 m. Calculate the pressure exerted on the mercury. **Answer: 10^{6} N. M^{-2}.**

10. Two vertical tubes of 2 cm^{2} connected by a horizontal tube contain mercury on the height of some centimeters. In one of the branches, 50 cm^{3} of a liquid is poured and the level of mercury drops by 1 cm. Calculate the weight by volume of the added liquid. **Answer: 1.088 g'. Cm^{-3}.**

6. Gas

Gas is a fluid in which the molecules exert a mutual attraction between them that produces a weak cohesion of very long distance. Gases exert an atmospheric pressure on the surface of any body encountered with a density :

$d = \frac{m}{29}$

where

d : density (no units),

m : mass in kilograms.

N.B.: The gas density is determined by a densimeter which corresponds to a float of particular shape.

Ref : J. Jodogne ; Physique 3, Chaleur ; A. De Boeck 1969, p135.

Application exercises

1. An oxygen occupies one volume under normal (standard) conditions of temperature and pressure. Calculate the density of oxygen.

Answer

Data	Request	Formula
M = 39 g	d = ?	$d = \frac{M}{29}$
Reasoning		
$d = \frac{39}{29}$ d = 1,34		

2. A compressed hydrogen in a closed vessel occupies a volume under standard conditions of pressure and temperature. Calculate the density of the hydrogen that exerts a useful pressure on the wall of the closed vessel. **Answer: 0.07**.

3. A carbon dioxide of 988 g occupies a volume of 500 cm ^{3}. Calculate the density of carbon dioxide. **Answer: 1.61**

4. An iceberg of 300 cm ^{3} floats on sea water with a density of 1.03 gf. cm $^{-3}$. Calculate the volume of water displaced by the iceberg whose density is 0.92 gf. cm $^{-3}$. **Answer: 268 cm ^{3}**

7. Submerged body

The immersed body is a solid totally immersed in a corresponding liquid. Immersed bodies have an apparent weight :

$P_a = P - P'$

where P_a : apparent weight in N or kgp

P: body weight in N or kgp

P' : thrust of the liquid in N or in kgp.

N.B.: Under emergence conditions, the body is subjected to the action of two vertical forces in opposite directions; P>P' and $V' = \frac{P}{\bar{W}}$

Application exercises

1. A copper sphere with a density of 9 gf. cm $^{-3}$ weighs 2.5 kgp in a liquid of density 0.8 gf. cm $^{-3}$. Calculate the apparent weight of the copper sphere immersed in the liquid.

Answer

Data	Request	Formula
$\bar{W}_s = 9\ gf.cm^{-3}$ $P = 2.5\ kgp$ $\bar{W}_e = 0{,}8\ gf.cm^{-3}$	$P_a = ?$ in gf	$P_a = P - P'$
Reasoning		
$P_a = P - \bar{W}V' = 2500 - 0{,}8\frac{2500}{9}$ $P_a = 2500 - 250$ $P_a = 2250\ gf = 22.1\ N$		

2. A copper sphere weighing 2.5 kgp is immersed in a liquid of density 1.68 gf. cm $^{-3.}$ Calculate the apparent weight of the copper sphere, whose density is 9gf. cm $^{-3}$. **Answer : 1975 gf**

8. Emerged body

The emerged body is a solid that floats on the surface of a corresponding liquid. The emerged bodies are particularly immersed in a liquid which develops a thrust given by :

$$P' = \bar{W}V'$$

where P' : thrust of a liquid in kgf

$\bar{W}$: Weight by volume in kg. M $^{-3}$

V' : volume of the liquid in m ^{3}.

N.B.: In this case; P = P'.

Under emergence conditions; P < P' whose emergent volume V e = V - V' and body volume is

.

The density of water is 1 gf. cm $^{-3}$.

Ref : J. Jodogne ; Physique 1, mécanique ; A De Boeck 1978, p166.

Application exercises

1. A block of wood with a density of 0.8 kgf. Cm $^{-3}$ weighs 5000 gp on water. Calculate the emergent volume of a block of wood floating on water.

Answer

Data	Request	Formula
P = 5000 gp $\bar{W}$ = 0.8 Kgf. cm $^{-3}$	V_e = ? in cm3	$V_e = V - V'$
Reasoning		
P = P'; 5000 = 1 V' V' = 5000 cm ^{3}; $V = \frac{5000}{0,8} = 6250\,cm^3$ V_e = 6250 - 5000 = 1250 cm ^{3}		

2. An iceberg of 300cm ^{3} floats on sea water whose density is 1.03 gf. cm $^{-3}$. Calculate the volume of water displaced by the iceberg whose density is 0.92gf. cm $^{-3}$.

Answer: 268 cm3.

CHAPTER V: CONDUCTIVITY

Objectives

Knowledge: give the definition of a molecule

Know-how: give the characteristics of the particles that make up an atom

A. Molecular structure

1. Molecule

The molecule is a particle of a pure body that can exist in a free state. The molecules are composed of atoms. The molecules are separated from each other by intermolecular pores that exist as empty spaces.

In solids, the molecules are stacked on top of each other in a regular and fixed order. They exert between them a mutual attraction which produces a strong cohesion of small distance.

In liquids, the molecules are piled up one on top of the other in a regular order. They exert between them a mutual attraction which produces a weak cohesion of medium distance.

In gases, the molecules are distant one on the other without regular order. They produce an elementary chao by shock.

N.B.: The distance and the agitation of the molecules are at the base of the difference of physical state of the matter.

2. Atom

The atom is a particle of a pure body that can enter into combination. Atoms are made up of nucleus and electrons.

Atoms are fused in chain synthesis reactions in a nucleon reactor. Atoms are fissioned during chain decomposition reaction in a nuclear reactor by neutron. Atoms are transformed into positive ions when they give up electrons by default and they are transformed into negative ions when they capture electrons by excess.

Atoms are transformed into ions by the transfer of electrons from one atom to another. Atoms are transformed into proton by ionization of an atom deprived of its electron.

The atoms that give up electrons are called cations and those that take up electrons are called anions.

N.B. : In simple bodies (iron, copper, oxygen, hydrogen, ..), the atoms keep the same properties as the molecule and are identical between them. In compound bodies (water, benzene, sodium chloride, ...), the atoms are different from each other in the molecule. In the molecule, the atoms are present in quantity and not in number.

3. Core

The nucleus is a fundamental particle that concentrates all the mass of an atom. Nuclei exist as the core of an atom.

Nuclei are composed of the proton, which exists as a primordial particle, and the neutron. Nuclei release neutrons which are transformed into protons by explosion in a detonator.

Nuclei are fissioned in a chain reaction in a nuclear reactor which provides a huge amount of release energy. Nuclei are fused in a chain reaction in an atomic reactor which provides a huge amount of cohesive energy.

Nuclei spontaneously emit triple radiation by radioactive decay. The nuclei release the proton by influence of x-particles endowed with a great energy.

The nuclei emit by radioactive decay beta radiation (β) that curves towards the positive plateau, gamma radiation (γ) that goes without deviation and alpha radiation (α) that curves towards the negative plateau.

N.B.: Gamma rays are similar to x-rays of light. Beta radiation is similar to cathode rays. The atomic mass of the nucleus is 2.0166 amu (1 amu = 1.666×10^{-27} kg).

4. Electron

The electron is a particle that is found on a peripheral layer of an atom. The electrons orbit around the nucleus by a revolutionary motion. Electrons that orbit the nucleus are called planetary electrons. The electrons repel each other on all the surface offered to them. Electrons move from one orbit to another by ionization of an atom.

Electrons that move in a conductor are called free electrons. The electrons remain localized in an electrical insulator, notably called dielectric.

The electrons remain localized at the place where they are produced in the insulator. The electrons are present in some number in the atom. The electrons are determined by an electrometer installed in an electric circuit.

N.B.: The mass of an electron is 9×10^{-31} kg,

The diameter of an electron is 3×10^{-15} m,

The radius of an electron is 2.8×10^{-13} cm.

B. Action of the electric current

1. Electric charge

Electric charge is a finite amount of electricity caused by the electrification of bodies. Electric charges correspond to electrons and protons that are in equilibrium in an atom where electrical work is performed.

Electric charges exert an electric force. Electric charges with the same signs repel each other with a force of repulsion. Electric charges with opposite signs attract each other with a force of attraction.

Electrical charges are detected by an electroscope. Electrical charges are measured in coulombs (C) as the primary unit in the International System (SI).

N.B.: 1 milli coulomb (mC) = 10^{-3} C

1 micro coulomb (μC) = 10^{-6} C

1 nano coulomb (nC) = 10^{-9C}

1 pico coulomb (pC) = 10^{-12C}.

2. Electrified body

The electrified body is a reference particle that carries a charge relative to the charge of the nucleus. Electrified bodies are charged with negative electricity by contact due to an excess of electrons.

Electrified bodies are charged with positive electricity by influence due to a deficit of electrons. Electrified bodies are returned to the neutral state by grounding because of an exchange of electrons. The bodies electrified by shock emit light.

Frictional electrified bodies attract metals. Electrified bodies create an electric field where the electrical effects of the current are felt, including: magnetic effects, mechanical effects and chemical effects.

Electrified bodies charged with negative electricity correspond to electrons and negative ions and those charged with positive electricity correspond to protons and positive ions.

N.B.: The elementary charge of an electron is -1.6×10^{-19} C ;

The elementary charge of a proton is 1.6×10^{-19} C ;

The state of the electrified bodies is detected by an electroscope.

3. Electric current intensity.

The intensity of the electric current is a characteristic of the electric current which corresponds to the displacement of the conduction electrons. The intensity of the electric current is accompanied by a quantity of electricity:

Q = It

Where Q : quantity of electricity in Coulombs,

I : current intensity in Amperes,

t: time in seconds.

N.B.: The quantity of electricity corresponds to the capacity of a generator; in particular 1 Ah = 360^{0} C, the quantity of electricity consumed is indicated by a meter. The intensity of the electric current is measured by an ammeter.

Exercises on current intensity

1. An electric current of 10 A flows through a conductor in 1 hour. Calculate the electric charge that accompanies the electric current in the circuit.

Answer

Data	Request	Formula
I = 10 A t = 1 h	Q = ? in C	Q = It
Reasoning		
Q = 10 A x 3600 s Q = 36000 C = 10 Ah		

2. An electric current of 20 A is accompanied by a quantity of electricity in a circuit for 10 h. Calculate the quantity of electricity accompanying the electric current. **Answer: 200 Ah.**

3. An electric current of 20 A delivers a quantity of electricity of 7200 C. Calculate the time after which the electric current flows. **Answer: 1 hour.**

4. An electric current of 8 A is accompanied by an amount of electricity of 86 400 C in a conductor. Calculate the time during which the electric current passes through the conductor. **Answer: 3 h.**

5. A battery of 80 Ah delivers an electric current during 5 whole days. Calculate the intensity of the electric current. **Answer: 0.6 A**.

6. A 60 Ah battery is charged in 20 hours. Calculate the intensity of the charging current. **Answer: 3 A.**

Ref : J. Jodogne ; Physique 3 ième, électricité ; A De Boeck 1970, p165

4. Conduction electron

The conduction electron is a particle that moves in a conductor. Conduction electrons carry a quantity of electricity :

Q = n e

where Q : quantity of electricity in Coulombs,

n : number of electrons ,

e : charge of the electron (e = $1{,}6.10^{-19}$ C).

N.B. : The elementary charge of an electron corresponds to the elementary charge of an ion.

Ref : J. Jodogne ; physique 4, électricité ; A De Boeck 1970, p 82

Application exercises

1. In a conducting wire, 10^{19} electrons per second pass. Calculate the intensity of the electric current flowing in the wire.

Answer

Data	Request	Formula
n = 10^{19} electrons t = 1 s	I = ? in A	Q = n e
Reasoning		
Q = n e = It ; $I = \frac{n\,e}{t}$ $I = \frac{10^{19} \times 1{,}6.10^{-19}}{1}$ I = 1.6 A		

2. In an electric circuit 375×10^{20} electrons pass in 20 minutes. Calculate the intensity of the electric current. **Answer: 5 A.**

3. An electric current of 2 A flows through a conducting wire in 2 minutes. Calculate the number of electrons that pass through the wire.

Answer: 15×10^{19} electrons.

4. In a point of the electric circuit is evaluated 75×10^{22} electrons to the passage of the electric current. **Answer: 3 h 20 min.**

5. In a high-powered electron tube, the filament emits 3.4^{17} negatons in one second on a circuit on the plate. Calculate the electric current shown on the plate circuit. **Answer: 4.8×10^{-2} A.**

5. Direction of the electric current

The direction of electric current is an orientation of electric current from one terminal to another of a generator in an electric circuit. The direction of electric current is controlled from the negative terminal to the positive terminal in an internal circuit of the generator.

The direction of the electric current is controlled from the positive terminal to the negative terminal in the external circuit of the generator. The direction of the electric current is controlled by a generator that accumulates electrons at the negative terminal and removes them at the positive terminal.

The direction of the electric current in the external circuit is from the positive terminal to the negative terminal of the generator and the direction of the electric current in the internal circuit is from the negative terminal to the positive terminal of the generator.

N.B.: The electric current is initiated in the circuit when the direction of movement of the electrons goes from the positive terminal to the negative terminal in the generator and from the negative terminal to the positive terminal in the external circuit.

C. Properties of the material

Knowledge: give the properties of matter

Know-how: speak briefly

1. Electrical circuit

The electric circuit is a line of electric current which includes at least a generator, a conductor and a switch. The electric circuit is powered by a generator which supplies electric current across the resistance of a conductor.

The electric circuit is interrupted from the electric current by a switch. The electric circuit is closed when the electric current flows through it and is open when it is interrupted from the electric current. The electric circuit gives off heat dispersed in energy by Joule effect.

In an electric circuit, the electric current produces thermal (calorific) effects in heating devices, including: stove, stove, electric oven, iron, soldering iron, electric wire, electric bulbs, water heater, electric radiator, etc.

In an electric circuit, the electric current produces the light effects in lighting devices, including the electric bulb, incandescent tube, projector, etc.

In an electric circuit, the electric current produces physiological (biological) effects in the human body; namely burning, muscle contraction, blood clotting; in short electrocution.

In an electric circuit, the electric current produces chemical effects in chemical generators, namely: the battery, the cell; in short, the accumulators.

In an electric circuit, the electric current produces magnetic effects in mechanical generators, such as magnetos, alternators, and dynamos, and in measuring devices, such as voltmeters, ammeters, parallel current balances, compasses, electric meters, multimeters, and tensiometers, as well as in electrical safety devices, such as fuses, circuit breakers, voltage boosters, and transformers.

2. Generator

The generator is a source of the electric current absorbed in a circuit. Generators convert one form of energy into electrical energy in a circuit. Mechanical generators convert mechanical energy into electrical energy in the circuit.

The mechanical generators produce an alternating current in a circuit, in particular: the magneto, the alternator, except the dynamo which produces a variable direct current.

Chemical generators convert chemical energy into electrical energy in the circuit. Chemical generators produce a direct current, such as a battery.

Generators accumulate electrons at the negative terminal and remove electrons at the positive terminal in the electrical circuit. Generators create an excess of electrons at the negative terminal and a deficit of electrons at the positive terminal in an electrical circuit.

N.B.: The magneto is used for lighting bicycles and for the production of spark plugs in internal combustion engines. The alternator is used for the transmission of mains current. The dynamo is used for the production of more intense electric current.

3. Lead wire

The conductive wire is an electric cable that carries electric current by conduction in a circuit. The conductive wires allow the electric current to pass through them perfectly, which gives off heat. The conductive wires give off the heat absorbed by joule effects when the electric current flows through them.

Electrically conductive wires are also called heat conductive wires. Conductors are insulated from the environment by insulation. The wires attached to the pylons carry a mains current used for domestic installations.

N.B.: The insulation of a conductor wire is achieved by a thermal insulator which corresponds to the electrical insulator also called dielectric, including: glass, rubber, dry gas, ebonite, sulfur, porcelain, silk, kerosene, oil, etc..

Insulators are electrified by friction to generate electricity.

4. Electrolyte

The electrolyte is a liquid conductor of the electric current linked to a double migration of ions. The electrolytes allow themselves to be crossed by an electric current which produces a chemical decomposition called electrolysis.

Electrolytes are broken down by an electric current that forms ions on the electrodes. Electrolytes are acids, bases, solutions, etc.

N.B.: The electrode is a metal immersed in the electrolyte. The electrode connected to the positive terminal of the generator is called anode and the one connected to the negative terminal of the generator is called cathode. Electrolysis is a chemical decomposition produced by the electric current which forms a metal at the cathode and releases a non-metal (gas) at the anode.

5. Rectifier

The rectifier is a transformer of alternating current into direct current in an electronic circuit. Rectifiers are also used to charge batteries. The simple alternating rectifiers (diodes) let pass alternating current in only one direction by suppressing one of the two alternations.

The simple alternating rectifiers called diodes, are classified into two categories, including lifetime diodes and junction diodes.

Full wave rectifiers called diode bridges rectify both waves of the alternating current. Full wave rectifiers combined with capacitors improve the rectification of the alternating current and allow to obtain a filtered (boosted) current similar to a current supplied by a battery.

Rectifiers are used in televisions, telephones, radios, computers, radars, etc.

N.B.: Radio and television receivers are powered by the accumulators that supply electric current.

6. Capacitor

The capacitor is a conservator of electrical charge carried to the electrical potentials of a collector. Capacitors are composed of two conductive plates called armatures, separated by an insulating material called dielectric (Fondanèche, P et al., 1966, page 96).

The capacitors are mounted in a circuit comprising a generator, an ammeter to measure the intensity of the electric current and a voltmeter to measure the voltage across the armatures.

Capacitors are charged to maximum voltage when the potentials of conducting plates and terminals of a generator are equal; therefore, there is no more electric current.

Capacitors are discharged under breakdown voltage when there is a potential difference (ddp) between the positively and negatively charged armatures; therefore, there is electric current flowing from positive to negative potential. Capacitors are classified into two categories; namely fixed (planar) capacitor and variable capacitor.

N.B.: In a discharged capacitor, the electric current will be established from the positive potential to the negative potential when the surplus electrons will flow from the negative plate to the positive plate where the electron deficiency will be compensated.

In a charged capacitor, the electric current no longer flows because the potentials of the conducting plates and the generator terminals are equal.

The discharge of a capacitor will take longer when the capacitance of the capacitor is larger. The electric current is established in equilibrium from the negative plate to the positive plate of a capacitor by a potential difference ***(pd)***.

7. Transistor

The transistor is a semiconductor crystal that participates in the conduction from the ordinary temperature in an electric circuit. The transistors formed by atoms having a valence electron more (phosphorus, arsenic, ...) have a high concentration of carriers similar to negative charges. The transistors formed by atoms with one less valence electron (Aluminum) have an important concentration of holes similar to positive charges.

Transistors made of foreign atoms work as n-type doped electron donors and p-type doped electron acceptors. The transistors work as an amplifier in an oscillating circuit that includes a pure resistor and a coil of conducting wire.

Transistors work as an oscillator in an oscillating circuit comprising a coil and a capacitor.

Transistors produce continuous electrical oscillations of constant amplitude like a triode. Transistors work in the same way as a triode. Transistors are used in electronic devices.

Invention of the transistor

American physicists Walter H. Brattain, John Bardeen and William B. Schocley develop the transistor, an electronic device that uses the semiconducting properties of certain materials to amplify signals. Replacing vacuum tubes, the transistor became the basic element of modern electronics.

8. Microphone

The microphone is an acoustic transmitter that transmits modulated electrical current to make the telephone talk.

The microphone transforms the mechanical vibrations of sound into modulated electrical vibrations. The microphone transforms mechanical energy into modulated electrical energy in conducting wires. The microphone transforms the biological energy of the person making the sound into mechanical energy in the air.

The microphone transmits modulated electric current through the conductor wires to make the telephone talk. The microphone is operated by an electric current which creates a magnetic field in the turns of the wire coil.

The microphone has an electromagnet that causes a soft iron membrane to vibrate with an electric current.

N.B.: The soft iron membrane moves according to all the vibrations modulated by the electric current.

The modulated vibrations correspond to the frequency of the microphonic current. The magnetic field is the region of space where the magnetic effects of the electric current are felt.

9. Phone

The telephone is an acoustic earpiece that reproduces exactly the sound following all the modulated vibrations of the voice. The telephone allows you to listen to the voice of the person making the sound on the cellular device.

The telephone produces a reversible transformation of modulated vibrations into sound vibrations. The telephone operates by an electric current that creates a magnetic field in the turns of the conductive wire coil. The telephone is used for immediate voice communication of information in offices and throughout the modern world.

The telephone produces a reversible transformation of radio waves in antennas, especially transmitting and receiving antennas.

The telephone includes an electromagnet which makes a soft iron membrane vibrate by influence of the magnetic field. The transmission telephone includes a generator of continuous radio waves, a transmitter and a modulator.

The telephone receiver includes an antenna, an electromagnetic wave detector, an amplifier and a speaker.

The transmitting and receiving phones are separated by thousands of miles.

A conventional telephone set consists of the following components: a transmitter or microphone, a receiver or earpiece, a ringer system, a keypad and an echo suppressor. The receiver and microphone are located in the handset, the ringer is integrated into the base, and the keypad and voice coil are installed in the base or handset. In addition, the base usually contains a speaker and a microphone in addition to the handset, which allows for amplified listening or even hands-free use. On a cordless phone, the handset cord is replaced by a radio link between the handset and the base.
N.B.: The soft iron membrane vibrates due to the influence of the magnetic field which exerts a carrying force on the hammer. Hertzian waves are electromagnetic waves used for communication and information between people at a distance of thousands of kilometers.

By inserting the microphone and the telephone in one book, the telephone number of the person requested can be formed. The ringing tone announces a formation of the requested person's telephone number.

BIBLIOGRAPHIC REFERENCES

- Basquin, R. (1960): Mécanique expérimentale; Paris.
- Bribosia, A (1992) : physique 5 ième, option complémentaire ; A De Boeck.
- Faucher, R. (1966):
- Fondanèche, P and Roulet, B. (1966): Electricité, Tome 2, Nathan technique, 18, Rue Monsieur le Prince-Paris VIe.
- Fontaine, G. (19):
- Guide in support of the Programme Educatif du Domaine d'Apprentissage des Sciences. Classe de 1 ière année des Humanités Scientifiques; 1 st edition, Kinshasa, 2019.
- Jodogne, J. (1970) : Physique 3 ième, électricité ; A De Boeck.
- Jodogne, J. (1969): Physique 3, chaleur; A De Boeck.
- Jodogne, J. (1970) : Physique 3 ième ; A De Boeck.
- Jodogne J. (1970): Physique 4, électricité; A De Boeck.
- Jodogne J. (1978): Physique 1, mécanique; A De Boeck.
- LISMONT, I and DEWALLER, J: Applied Physics, Haute école de la Province de Liège. Agronomy category, Haut Marêt 20. 4910. Bachelor in Agronomy.
- Mpanda, M and Luyindula, M. (1987): Exercices de physique, Mécanique, edition Centre de recherche pédagogique. B.P. 1900, Kinshasa I
- Microsoft ® Encarta ® 2009. 1993-2008 Microsoft Corporation. Accessed on Wednesday, February 17, 2021 at 11:45 AM.
- PIRA, E. (1973): Energy; Kinshasa.
- Possoz J.; Technologie 2 ième année secondaire ; Kinshasa, 1989.
- Educational program of the Science Learning Domain. Class of 1st year of Scientific Humanities, 1st edition, Kinshasa 2019.
- Smolders, P; (1985): Exercices résolus de physique, enseignement général, C.R.P. Kinshasa.

Printed by Books on Demand GmbH, Norderstedt / Germany